BarCharts, Inc.®
WORLD'S #1 ACADEMIC OUTLINE
W9-BXG-331
Quick Study® ACADEMIC

ENVIRONMENTAL SCIENCE

Relationships of Humans to Their Environment

LIFE ON EARTH

Earth – only known location of life in the Universe

1. **Biosphere:** Where life exists on Earth
2. **Geology:** Study of the physical Earth
   a. Consolidated materials – rocks
   b. Unconsolidated materials – sediments and soils
   c. Earth's Age: 4.5 billion years, sequentially divided into the following time frames: Precambrian → Paleozoic Era → Mesozoic Era → Cenozoic Era
3. **Water (Hydrosphere):** Covers 70% of Earth's surface
   a. Distribution: 97% seawater; 3% freshwater (2% in ice; 1% in lakes, rivers and atmosphere)
4. **Atmosphere:** Gases held by Earth's gravitational forces
   a. Composition: 78% nitrogen; 21% oxygen; <1% carbon dioxide; <1% water vapor (although near surface, may vary 1–4%)

History of Humans & Sustainability

1. Earliest humans were likely hunters/gatherers
2. Eventually (approx. 10,000 years ago), we domesticated animals and plants, triggering an **agricultural revolution**
3. With sufficient food and resources, people have lived longer and reproduced more, triggering population growth
4. For most of our history, only millions of humans inhabited Earth
5. **Industrial revolution** helped trigger a major population increase during the mid 1700s
6. Currently, 1 million+ new humans are born every 4 days
7. Human population may reach 10–12 billion by 2050
8. Earth cannot "sustain" an unlimited number of humans (or other organisms) indefinitely
9. **Carrying capacity** of Earth for humans may be 10–20 billion

Natural Resources

1. Materials that could be used (i.e., **potential resources**) or are used (i.e., **actual resources**), and are derived from the environment
   a. **Biotic resources:** Obtained from organisms, including fossil fuels
   b. **Abiotic resources:** Obtained from non-living materials
   c. **Renewable resources:** Replenished in a "relatively" short time period (e.g., crops, forests) or are essentially unlimited (e.g., sunlight, wind, air, waves)
   d. **Non-renewable resources:** Created over geological time frames and are used faster than they can be replenished (e.g., fossil fuels), but some can be **recycled** (e.g., minerals, precious metals)
2. **Natural Resource Management**
   a. Management of natural resources (biotic and abiotic) for human welfare; includes **sustainable development,** which involves management of land use

Environmental Science vs. Environmentalism vs. Ecology

1. **Environmental Science:** Focuses on studying relationships of humans to our environment
2. **Environmentalism:** A philosophical and social approach dedicated to protecting the Earth from human-induced negative changes
3. **Ecology:** Focuses on relationships of organisms to their environment

Ecological concepts apply to environmental science

1. **Ecosystems:** Interacting unit of biotic communities and abiotic environments
   a. **Trophic Structure**

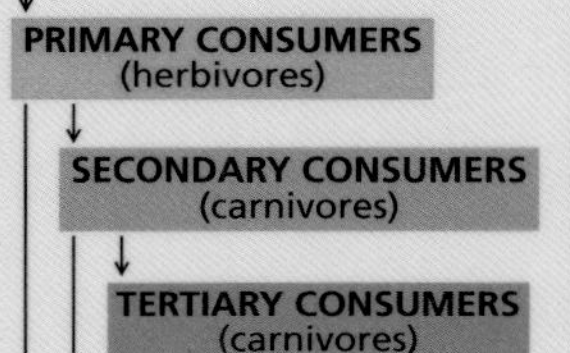

   - autotrophs – use either photosynthesis or chemosynthesis to produce food
   - heterotrophs – use food molecules produced by autotrophs

   b. **Major Components**
   - abiotic portions
     - sunlight (which drives most systems)
     - inorganic and organic substances
     - environment
   - biotic portions
     - producers (autotrophs)
     - consumers (heterotrophs), including humans
     - decomposers recycle organic debris

2. **Energy Flow**
   a. **Laws of Thermodynamics**
   - **1st Law – Conservation of Energy:** Energy can neither be created nor destroyed, but can be converted from one form or another
   - **2nd Law – Law of Entropy:** In any energy conversion, less usable energy will be available after each conversion due to heat loss; matter tends to become less organized and more random (a condition called *entropy*)

   b. **Energy Pyramids**

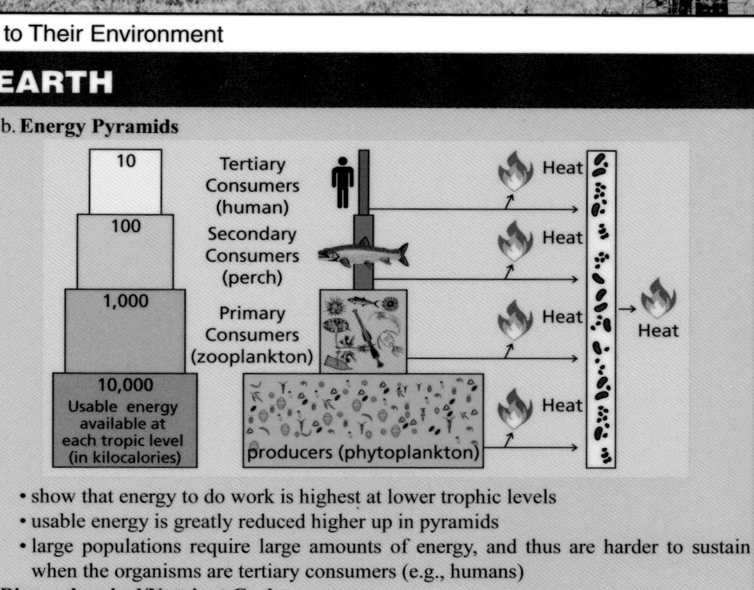

   - show that energy to do work is highest at lower trophic levels
   - usable energy is greatly reduced higher up in pyramids
   - large populations require large amounts of energy, and thus are harder to sustain when the organisms are tertiary consumers (e.g., humans)

3. **Biogeochemical/Nutrient Cycles**
   a. *"Message from Nature – **Recycle**"*
   b. Most minerals/nutrients move between four compartments in nature

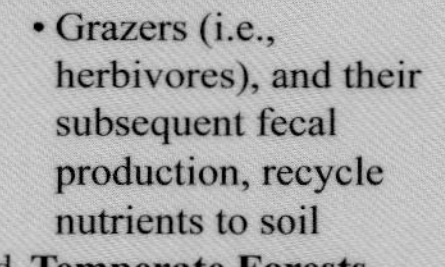

   c. **Grasslands**
   - Grazers (i.e., herbivores), and their subsequent fecal production, recycle nutrients to soil

   d. **Temperate Forests**
   - nutrient reservoir is primarily in the soil, which is subjected to little rainfall compared to tropical rain forests; nutrient cycling is slower and steady nutrient levels are easier to maintain

   e. **Tropical Rain Forests**
   - nutrient reservoir is primarily in the biomass of the plants, which must quickly take up nutrients before they are washed away by rainfall

   f. **Deserts**
   - extremely harsh environment where primary productivity is constrained by very low rainfall; little decomposition, extremely low nutrient cycling rates

   g. **Oceans**
   - mostly nutrient-impoverished environment; however, because of its large volume, nutrient cycling and primary productivity is significant to the biosphere
   - neritic (near the continents) areas, estuaries, mangrove forests, coral reefs, and hydrothermal vents are nutrient-enriched areas

   h. Three Specific Cycles – Water, Carbon, Nitrogen
   - **Water Cycle**
     - more water evaporates from the sea than returns to it directly by rainfall
     - less water evaporates from land than returns to it directly by rainfall
     - water mainly in balance between land and sea

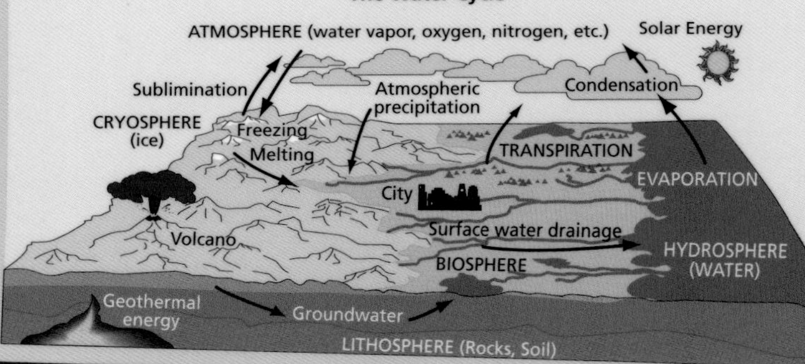

- **Carbon Cycle**

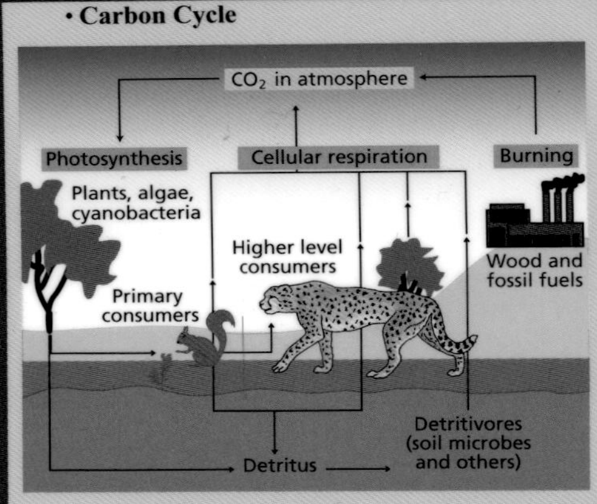

- CO_2 in the atmosphere and oceans is cycled relatively quickly
- photosynthesis uses CO_2
- respiration, combustion and deforestation produce CO_2
- carbon distribution in tropics versus temperate terrestrial regions:
  - * tropics: >75% in vegetation; <25% in soil
  - * temperate: <50% in vegetation; >50% in soil
- **Nitrogen Cycle**
  - N_2 gas makes up nearly 80% of air
  - most organisms can only use NH_3, NO_2, NO_3
  - N_2 is converted/fixed into usable form by organisms in two ways:
    - * electro- or photo-chemically
    - * biologically, using the special enzyme nitrogenase (bacteria, cyanobacteria and fungi)
  - bacteria are also involved in cycling nitrogen back to N_2

4. **Community:** Assemblage of interacting organisms in one environment
   a. Communities vary in species composition
   b. **Species dominance:** Numerical abundance is not the only criterion necessary for such designation
   c. **Keystone species:** Organisms that play a critical role in maintaining the integrity of the community; in undisturbed, natural communities it is not always easy to determine, but humans have become keystone species in many ecosystems
   d. **Indicator species:** Organisms that are sensitive to environmental change, such as a disease outbreak, pollution or climate change
   e. **Biodiversity & Evolution**
      - **Species Diversity = Biodiversity**
        - **species richness** – number of species, usually higher in complex habitats
        - **relative abundance** – number of individuals of one species compared to total for all species
      - evolution and natural selection have resulted in varying characteristics of species, which help define community structure
   f. **Ecological succession:** Changes in the composition and function of communities
      - **pioneer species** – early successional species
      - **primary succession** – community development on a site previously unoccupied (e.g., volcanic rock)
      - **secondary succession** – community development on a site previously occupied by a community (e.g., disturbance, such as fire, has cleared a vegetated area)
      - **climax community** – some communities reach maturity, and subsequently change little in species diversity unless major disturbance occurs

5. **Populations**

POPULATION SIZE

GROWTH FACTORS (Biotic Potential)	DECREASE FACTORS (Environmental Resistance)
Abiotic	**Abiotic**
• Favorable light	• Too much or too little light
• Favorable temperature	• Temperature too high or low
• Favorable chemical environment (optimal level of critical nutrients)	• Unfavorable chemical environment (too much or too little of critical nutrients)
Biotic	**Biotic**
• High reproductive rate	• Low reproductive rate
• Generalized niche	• Specialized niche
• Adequate food supply	• Inadequate food supply
• Suitable habitat	• Unsuitable or destroyed habitat
• Ability to compete for resources	• Too many competitors
• Ability to hide from or defend against predators	• Insufficient ability to hide from or defend against predators
• Ability to resist diseases and parasites	• Inability to resist disease and parasites
• Ability to migrate and live in other habitats	• Inability to migrate and live in other habitats
• Ability to adapt to environmental change	• Inability to adapt to environmental change

a. Factors affecting population size
- **density** – number of individuals in a given area
- **dispersion patterns**
  - **uniform** – competition may force individuals to be relatively evenly spaced
  - **random** – rare, as resources are not usually random
  - **clumped** – most common, as resources are usually patchy (e.g., for humans, cities)
- **natality** – ability to increase population size (e.g., birthrate)
- **mortality** – death of individuals
- **dispersion** – movement into (immigration) or away from (emigration) an environment
- **biotic potential** – inherent ability for a species to be successful
- **carrying capacity** – maximum population size for specified area

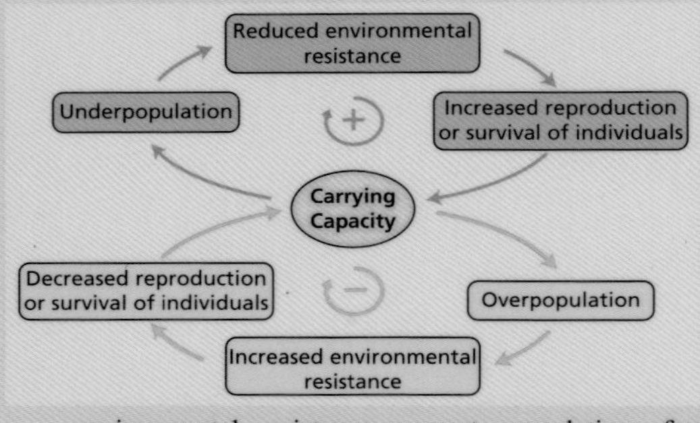

- environmental resistance prevents populations from increasing at their biotic potential indefinitely; thus, population sizes are limited within an ecosystem
- **growth form** – how population growth increases
  - two common mathematical curves are used to illustrate population growth

J-shaped & S-shaped Population Growth Curves

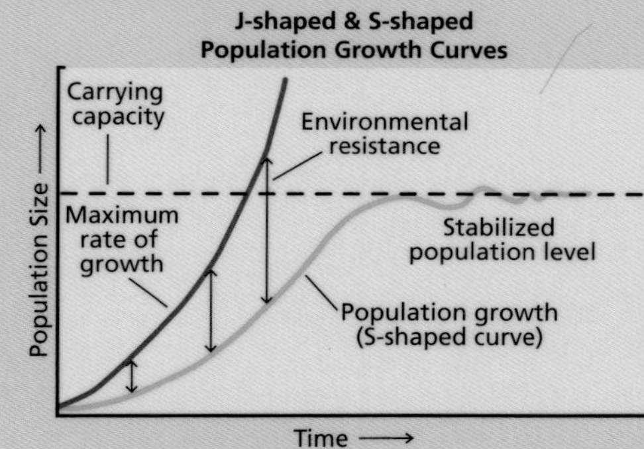

  - * **J-shaped curve:** Shows exponential growth, with maximum biotic potential and assuming no carrying capacity exists
  - * **S-shaped curve:** Shows logistic growth, factoring in the carrying capacity (i.e., environmental resistance)
- exponential growth can lead to population crashes if the carrying capacity is reached too quickly, in some cases causing dramatic decreases in the overall carrying capacity
- human population growth may quickly be approaching such a population crash

Population Crash from J-shaped Growth

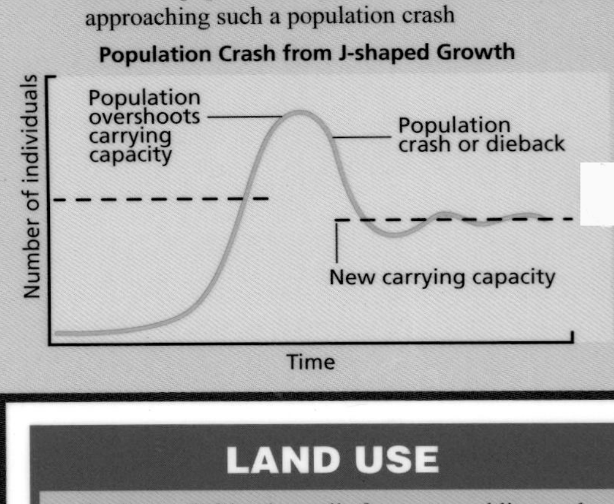

LAND USE

Agriculture: Cultivating soils for crops and livestock

1. **Agronomy:** Study of the use of plants primarily for food, feed and fuel, including environmental impact of agriculture
2. **Soil Science:** Study of soils as natural resources
   a. **soil** – degraded rock, organic matter, nutrients and microorganisms (could be defined as an ecosystem)
   b. **croplands** – soils that grow most of our food; 10–13% of land
   c. **rangelands** – soils that grow food for grazing livestock; 20–25% of land
3. **Soil Conservation & Degradation**
   a. ideal soils typically hold nutrients, have neutral pH, and are workable
   b. many soils are not ideal for agriculture
   c. **erosion** (via water or wind) can degrade soils by removing rich topsoil from an area
   d. **deposition** of eroded materials may enhance other areas, but typically the donor areas of soil are adversely impacted, as natural soil generation processes are much slower than erosion
   e. **desertification** – loss of productivity in arid or semi-arid areas, possibly leading to a desert
   f. human activities have degraded many ideal soils
      - **Dust Bowl** in the 1930s was triggered, in part, by early settlers overplowing for crops and overgrazing grasslands for their cattle, prompting Congress to pass *Soil Conservation Act of 1935*
   g. soil conservation methods
      - **crop rotation** – alternating crop types (e.g., wheat one year, then corn the next year) to help recycle nutrients, inhibit disease spread, and minimize erosion when lands are without crops
      - sloped lands are particularly vulnerable to erosion
        - **contouring** – digging grooves in the land to slow the effects of erosion

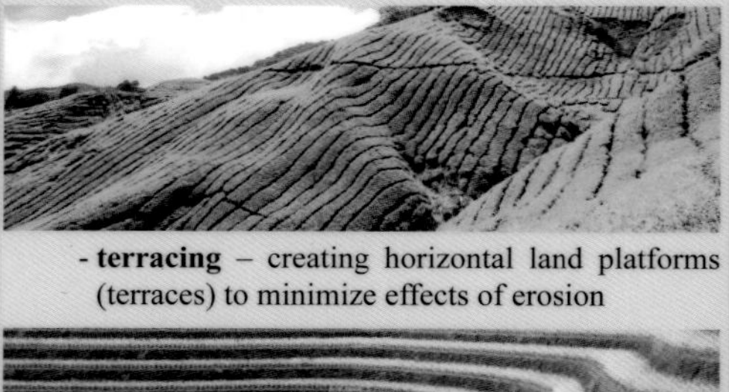

        - **terracing** – creating horizontal land platforms (terraces) to minimize effects of erosion

      - **shelterbelts (windbreaks)** – planted along edges of farm plots to deflect wind, thus minimizing erosion on croplands

4. **Fertilizers:** Essential nutrients (e.g., nitrogen and phosphorus) added to soils to enhance crop growth
   a. **organic fertilizers** – from natural sources, such as manure and compost
   b. **inorganic fertilizers** – synthesized or mined nutrients
   c. misuse/overuse can lead to ecological problems; organic fertilizers generally trigger fewer problems than inorganic fertilizers
5. Controlling crop pests (e.g., insects, plants, fungi)
   a. **pesticides** – synthesized chemicals, such as insecticides, herbicides and fungicides; may have serious environmental impacts; pests may also evolve to be resistant, triggering the need to synthesize new, more toxic pesticides
   b. **biological control** – using organisms to attack pests of crops; also have serious risks in that imported organisms may have wider, long-term ecological impacts
6. **Organic agriculture:** Congress passed the *Organic Food Production Act* in 1990, establishing standards for "organic foods" including the non-use of pesticides and synthetic fertilizers
7. **Wetlands conversion:** Most of the natural wetlands in the U.S. have been drained and converted into farmlands; we now realize the value of wetlands and have been more cautious and slowed the rate of conversions

Forests: 20-30% of land on Earth is covered by forests

1. **Deforestation:** Wood products (e.g., paper, building materials), farming and urbanization have triggered the clearing of vast stands of trees worldwide
   a. timber production has basically stabilized in the U.S., through managed cutting strategies
   b. deforestation to convert areas into croplands in developing countries is increasing (e.g., South American tropical rain forests)
   c. deforested tropical regions do not sustain agricultural crops for more than a few years before nutrients are depleted, as most nutrients in these ecosystems are stored in the plants

Annual net change in forest area by region (1990-2005)
million hectares per year

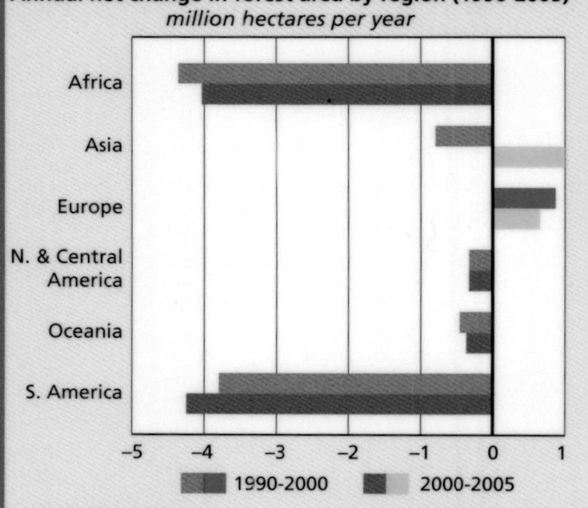

   d. deforested temperate regions can sustain crops for many years because most nutrients are locked in the soil; for this reason, most present-day human populations have developed in temperate regions that can "feed the people"
   e. loss of tropical rain forests leads to a loss of biodiversity of flora/fauna

Mining: Extraction of resources (minerals, geological materials) from the Earth, specifically including metals, uranium, coal, diamonds, limestone, shale, oil and natural gas

1. **Environmental Impacts**
   a. erosion
   b. contamination of soil and water
   c. loss of habitat and biodiversity
   d. abandoned mines can be dangerous, with toxic gases and compromised shafts

Parks & Reserves: Tracts of land (and sometimes water) set aside to be protected from development, based on their natural beauty, uniqueness and significance in an ecosystem

1. **Yellowstone National Park** – established in 1872
2. **National Park Service** – established in 1916 to administer parks, reserves and monuments
3. **U.S. Fish and Wildlife Service** – administers National Wildlife Refuges, which sometimes allow hunting and fishing, while maintaining biodiversity through restoration and reintroduction of species in decline
4. **City Parks** – similar to national parks, but on smaller scale

Urbanization: Development of urban population centers from undeveloped lands, including immigration of people from rural areas and/or conversion of rural areas into cities

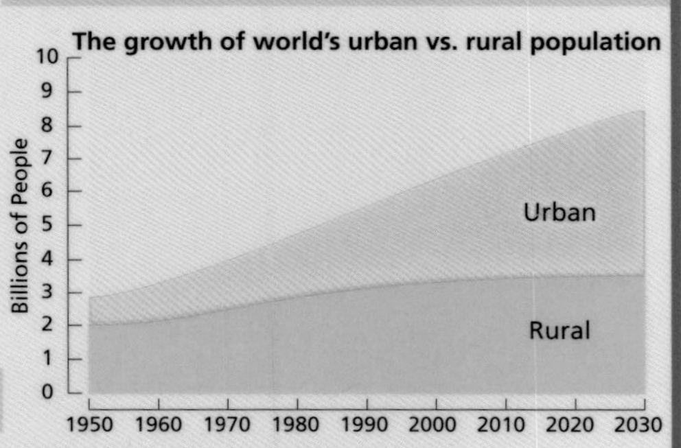

1. **Environmental Impacts**
   a. **heat island** – large cities can be warmer by 2–10° F (1–6° C) than surrounding rural areas because:

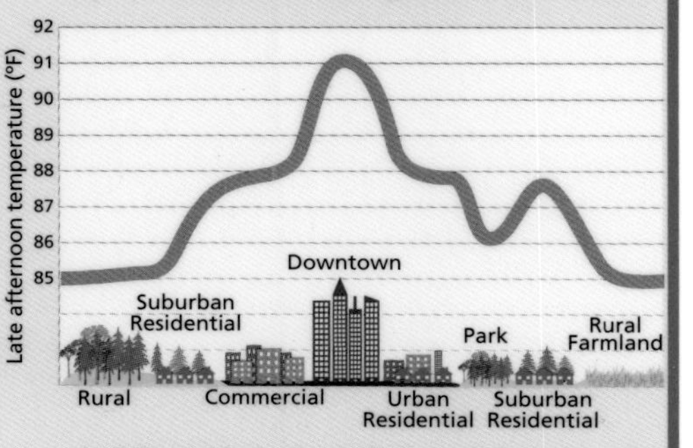

   - buildings block the radiation of heat into the atmosphere, especially at night
   - vegetation typically cools an area through evapotranspiration of water from leaf surfaces; urban areas have sparse vegetation; attempts to mitigate problem involve planting more vegetation (typically trees)
   - building materials (e.g., concrete and asphalt) typically have thermal properties conducive to heat retention (high heat capacity/low radiative reflectivity)

   b. **air pollution** – carbon dioxide (and other gas combustion and industrial byproducts) can accumulate to high levels, creating health issues for humans (e.g., smog)
   c. **flooding** – rural areas with natural soils can typically absorb high levels of rainfall; cities paved with concrete/asphalt can create flooding during heavy rains, as municipal drainage systems may be overwhelmed
   d. **transportation** – population increase and urbanization drives the need to create roads and vehicles to travel on these roads; specific impacts include:
   - land acquisition, which diminishes croplands, grazing lands and ecosystem habitats
   - increased energy consumption
   - pollution (air, water, land and noise)

   e. **urban planning** – attempts to integrate land use and transport planning to mitigate possible environmental impacts, while improving quality of life

WATER USE

All organisms metabolically need water; humans also use water for agricultural, household, industrial and recreational purposes

Most of our water "needs" involve freshwater, which, unfortunately, is in limited supply—3% of total water, as compared to the abundance of seawater

Additionally, most human population centers are near coastlines; thus, meeting freshwater demands becomes more challenging with seawater abutting natural freshwater supplies

1. **Freshwater Systems**
   a. **groundwater** – water from rainfall and rivers percolates into the soil, and accumulates in **aquifers** (composed of porous sediments, including rock, sand or gravel)
   - **aquifer recharge areas** – where water percolates into soil to continually "recharge" or supply aquifer with water
   - turnover rate can be slow; water in aquifers may be thousands of years old

   b. **freshwater ecosystems**
   - **limnology** – study of inland **surface waters**; primarily freshwater
     - **lentic systems** – have non-moving waters; lakes (large) and ponds (small)
     - **lotic systems** – have actively flowing waters; rivers and streams
     - **wetlands** – marshes and swamps; highly productive areas
   - freshwater ecosystems are important for recreational fishing

   c. **freshwater use**
   - **consumptive use** – water is removed from aquifers and surface waters, but is not returned directly to these sources; includes 70% used for agricultural irrigation; industrial and residential make up remaining use
   - **non-consumptive use** – does not remove, or removes only temporarily, water from its aquifer or surface-water source; includes hydroelectric dams
   - **flood control** – to avoid flooding of urban regions, typically large amounts of freshwater are diverted
   - **freshwater depletion**
     - **water "quantity" supply** – based on global consumption rate, we are depleting both groundwater and surface waters, necessitating in many regions the adoption of water restrictions (typically involving agricultural and residential irrigation)
     - **water "quality" supply** – as aquifer water tables decline, mineral concentrations can increase, especially near coastlines where **saltwater intrusion** can occur; wells pumping water out of these aquifers can exacerbate the problem, due to the vacuum pressure exerted on the aquifers
   - **increasing water supply** – although the total amount of water on Earth is basically constant, our demands have been ever increasing; several strategies can increase the supply of "usable" water:
     - reduce demand, primarily through conservation practices, such as efficient irrigation (e.g., drip irrigation); 50% of crop irrigation is lost through percolation and evaporation
     - the oceans hold most of the water, but the high salt content restricts the direct use of seawater; many arid regions or small-island nations can only sustain population water demands using **desalination** (desalinization) processes to remove the necessary salts for consumption; it is relatively expensive, especially for nations with large ground- or surface-water supplies, but as demand for water increases, this process may become a more common strategy worldwide

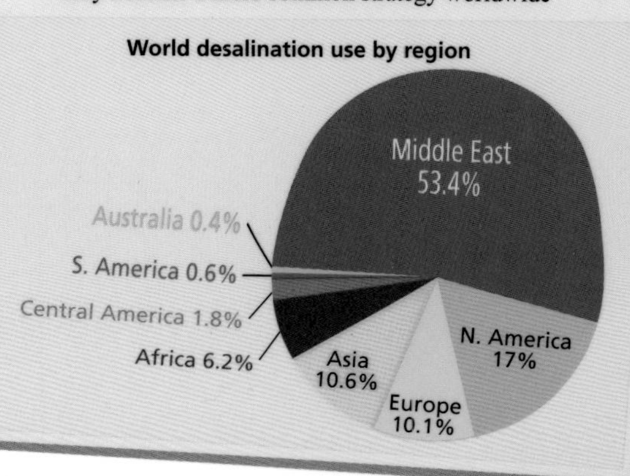

ENERGY RESOURCES

Energy is a requirement at all levels of matter, from abiotic to biotic systems; humans have harnessed various types of energy forms to enhance standard of living across the world; leading the way in energy development are "developed" countries; unfortunately, demand for inexpensive, vast amounts of energy typically exceeds resource supply of certain energy forms available within borders of developed countries (e.g., the U.S.)

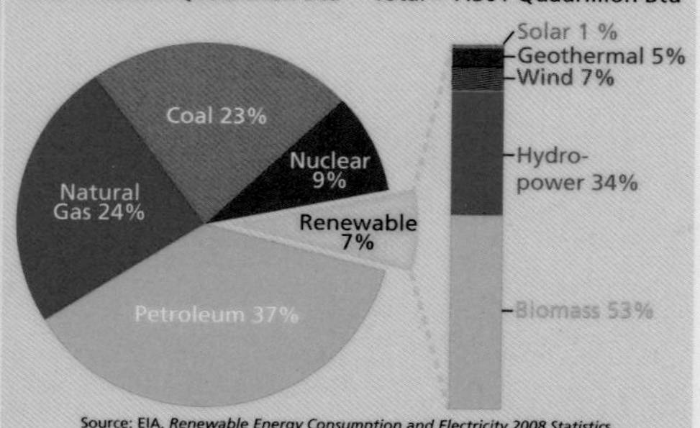

1. **Non-renewable energy** has been the leading source of energy consumption, but is based on relatively finite resources
   a. **fossil fuels** include three major forms: **oil, natural gas, coal**
      - in the U.S., more **gasoline** (from refined oil) is consumed than the U.S. oil industry can produce, which in some cases has been partly responsible for international policies; oil also is used for heating
      - **natural gas** is a growing source of fuel worldwide

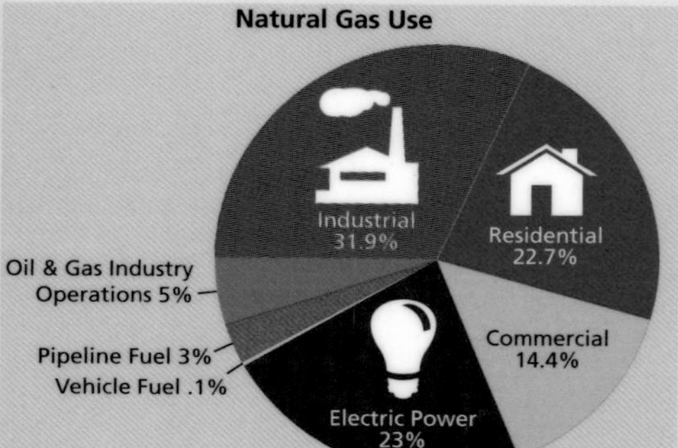

      - **coal** is the most abundant fossil fuel in the world, extracted via mining
   b. **nuclear power** comes from energy released during fission reactions involving uranium; U.S., France and Japan are leading nations in electricity generated from nuclear power plants; concerns about safety, based on major accidents (Three Mile Island, 1979; Chernobyl, 1986), and overall costs to build such plants, have slowed growth of nuclear power plants
2. Renewable energy is a growing source of consumption, based on "relatively" infinite resources
   a. **solar** – Sun presents a tremendous amount of potential energy to Earth; estimates have suggested that total solar energy striking our planet in a day would sustain human energy demands for nearly 30 years; 2 strategies exist to capture some of this energy:
      - **passive solar energy collection** – buildings, construction materials and strategies (e.g., window placement) are designed to absorb sunlight for heat production in winter, while maintaining cool conditions in summer
      - **active solar energy collection** – technology to capture and store solar energy for subsequent alternative uses (i.e., production of electricity)
   b. **hydrolelectric**
      - **freshwater** – damming of rivers worldwide has been used to have turbines turned by flowing, controlled water to generate electricity; some hydroelectric dams are huge (e.g., Hoover Dam on the Colorado River and Nurek Dam in Tajikistan)
      - **marine** – ocean currents and tides represent similar "potential" opportunities to river systems; major ocean currents (e.g., Gulf Stream) may also be used eventually
   c. **geothermal** – geological and chemical processes deep within Earth create a tremendous amount of heat that occasionally is transferred to the surface via magma; currently, geothermal energy is used to heat homes directly via **geothermal ground source heat pumps** (GSHPs) and to generate electricity by using steam generated from heated groundwater directly above heated rocks
   d. **biomass** – organic materials from living plants and animals can serve as "biofuels," such as wood, charcoal, urban yard waste, crops (both residual and primary components; e.g., cornstalks and corn, respectively), methane (from bacterial actions in landfills) and manure
   e. **wind** – wind energy, which is essentially generated from sunlight, can be harnessed by wind turbines to generate electricity—rapidly growing form of energy worldwide
   f. **hydrogen** – storing electricity, once generated from sources discussed above, is problematic; electricity can be used to produce hydrogen, which can then be used by fuel cells for various energetic needs, including powering vehicles
3. **Energy Conservation:** Clearly, energy demands of a growing human population can exceed resource capacities, especially considering that non-renewable fossil-fuel supplies will someday be exhausted; basically, energy conservation can occur by:
   a. **individual conservation** – reducing energy consumption in our homes/businesses (e.g., moderate thermostat settings for heating/cooling; reduced lighting/appliance usage)
   b. **societal conservation** – reducing energy consumption in our utilities and equipment production (e.g., use of **cogeneration** [reusing excess utility plant heat for home heating, as well as other kinds of energy conversion]; increased automobile gasoline efficiency [smaller cars, efficient engines, hybrid technology, etc.])

KEY GLOBAL ISSUES

Climate Change, Food Production & Hunger

1. **Climate Change**
   a. **Climatology:** Study of long-term (years to millennia) weather patterns (**meteorology** studies shorter-term temporal patterns), focusing primarily on the atmosphere, but also using data gathered from other disciplines, including geology and ocean sciences
   b. **Global Warming & Carbon Dioxide:** Geological data show that temperature fluctuations have occurred throughout Earth's history; we are now clearly in a warming trend, as glaciers have retreated to the poles from the last **Ice Age**; scientific measurements, however, indicate that **global warming** is occurring on a relatively rapid basis; when correlated to increases in carbon dioxide from anthropogenic inputs, most scientists agree that Earth's average temperature is accelerating, in part, due to human activity, as well as other atmospheric pollutants

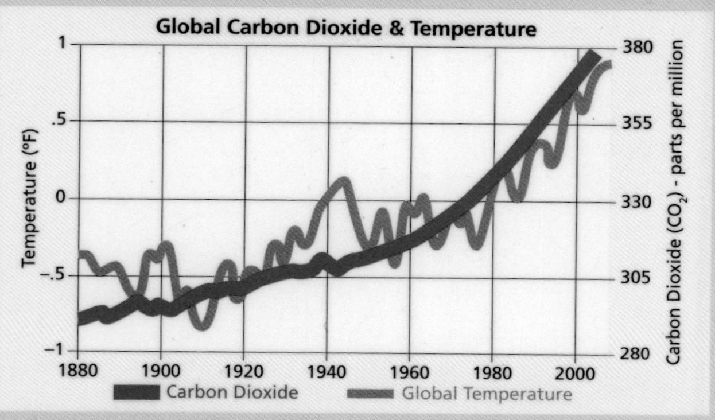

   c. **Intergovernmental Panel on Climate Change (IPCC):** An international group that assesses data, possible impacts and mitigation strategies of human-related global warming; IPCC's principal conclusions about **anthropogenic-induced global warming** include:
      - fossil-fuel emissions and climate change are linked
      - carbon dioxide levels are threatening marine fisheries and ecosystems through acidification
      - seawater temperature rise is accelerating, linking changes in both climate-level (e.g., El Niño and hurricanes) and ecosystem-level (e.g., coral bleaching events, biodiversity changes) phenomena
      - polar temperature rise in the Arctic and Antarctic, as well as sea ice melting, are accelerating, raising sea levels
      - global precipitation trends (i.e., dryer in some areas, wetter in others) are linked
   d. **Sustainable solutions to global warming**
      - because carbon dioxide is considered the most significant greenhouse gas, reducing emissions from vehicles and plants should be a high priority; however, many challenges lie ahead:
        - economic factors often outweigh environmental factors
        - responsibilities of industrial nations vs. developing nations
        - how to get nations to reduce emissions; voluntarily versus mandated
        - how to allocate funds and resources in order to reduce emissions
      - environmental movements, such as "green" actions, encourage individuals to get involved in resolving global climate change and to engage in practices that are ecologically healthy (e.g., less use of plastics, more recycling, etc.)
2. **World Food Production vs. World Hunger**
   a. Here are some sobering statistics regarding one of the greatest problems facing humans—hunger:
      - nearly 1 billion people are hungry (mostly women and children)
      - every 6 seconds, a child dies from hunger-related causes
      - more people die from hunger annually than die from diseases like AIDS and malaria
      - world food production (from agriculture, aquaculture, etc.) has increased
   b. A great paradox is that both hunger **AND** food production have increased worldwide; clearly, distribution of food is asymmetrical; developing countries are responsible for most recent population increases, while developed countries produce most of the food; this current situation presents economical, ethical, political and environmental issues—none of which can be easily solved

Author: W. Randy Brooks, PhD $6.95

ISBN-13: 978-142321425-0
ISBN-10: 142321425-0

2. **Marine Systems**

a. **marine ecosystems**

- **marine biology** and **oceanography** are key disciplines in the study of marine habitats, which are primarily categorized as:
  - **coastal zones** – estuaries, wetlands (including salt marshes, intertidal and coral reefs)
  - **pelagic** – open sea, water column
  - **benthic** – open sea, bottom; including hydrothermal vent communities

b. **ocean use**

- **transportation** – humans have used oceans for transport for much of recorded history; ships have sometimes inadvertently impacted the environment by introducing exotic species into new aquatic and terrestrial habitats; shipping also contributes to ocean pollution when oil cargo leaks
- **energy & minerals** – extraction of oil and natural gas from benthic oceanic habitats is significant worldwide; extraction of minerals (e.g., silica, calcium carbonate, sulfur and precious metals) occurs, but is sometimes logistically difficult to justify economically (e.g., manganese nodules)
- **fishing** – marine animals (fish, shellfish, etc.), along with agriculture, provide the bulk of the food supply for human populations; certain ocean phenomena, such as **coastal upwelling,** provide ideal conditions for tremendous aggregations of fish and other sea life; with technological advances (e.g., using radar, satellites, airplanes, state-of-the-art ships) in locating schools of fish, extremely efficient and successful fishing with nets, trawls and long-lines has occurred in oceans throughout the world; several negative impacts result from such high-tech fishing
- **bycatch** – in most cases, additional species that were not targeted find their way into these nets, etc.; some of these species may be endangered/protected (e.g., sea turtles, dolphins); much fishing is done in international waters, with laws enacted attempting to mitigate such bycatch by use of **Bycatch Reducing Devices** (BRD) and excluder devices (e.g., **Turtle Excluder Device** [TED]) incorporated into the collection nets

Bycatch % versus gear type

Bottom trawls 25.1%
Shrimp trawls 46.9.%
Purse seine 0.7%
Gillnet 1.2%
Longline 3.9%
Pot/trap 4.5%
Midwater trawl 5.1%
Dredge 5.3%
Hook & line 7.2%

- **overfishing** – the success of the fishing industry has been so great that "most" fishery stocks have been overexploited worldwide and are in danger of crashing; shark fishing, a relatively new fishery—partly to supply for a delicacy dish called "shark-fin soup"—has put the entire taxonomic group in danger

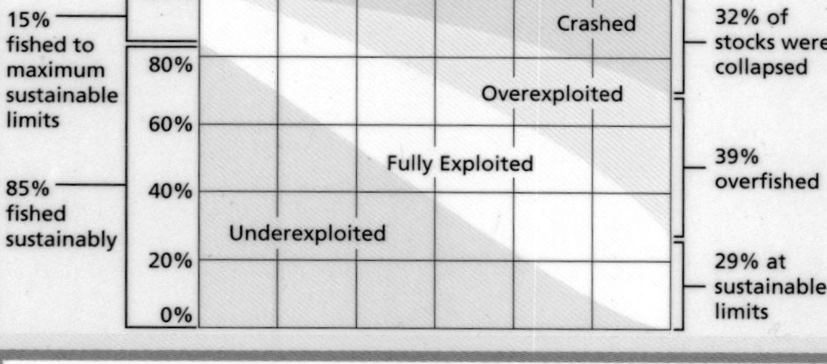

- **maximum sustainable yield** – fishing at the maximum level that will sustain future fishing success, is a current strategy employed; however, accurately calculating such levels has proven mostly elusive
- **marine protected areas** – while managed by governmental agencies, they do not typically prohibit fishing activities
- **marine reserves** – do not allow fishing, and thus, are considered by many as essential to maintaining ecosystem health in corresponding unprotected areas

POLLUTION

EPA: U.S. Environmental Protection Agency (U.S. EPA) is charged with protecting both human and environmental health through policy implementation (e.g., *Clean Air Act*) and enforcement

Air Pollution: Substances in the air that potentially harm humans and the environment

1. **Outdoor Air Pollution**

a. natural sources produce most air pollution, including volcanoes, forest fires, dust storms; human activities can exacerbate these processes (poor agricultural activities can lead to desertification and dust storms)

b. human activities can contribute to air pollution
- **stationary source** – refers to specific, unmoving spot (e.g., factory smokestacks)
- **mobile source** – refers to specific mobile sources (e.g., automobiles, airplanes, trains, ships)
- **primary pollutants** – are immediately harmful in their present form (e.g., carbon monoxide)
- **secondary pollutants** – are synthesized in the atmosphere from primary pollutants and natural compounds (e.g., sulfuric acid and ozone)

c. **EPA-regulated pollutants**

Air Quality Index (AQI Values)	Health Concern Level	Color Symbol
0 to 50	Good	Green
51 to 100	Moderate	Yellow
101 to 150	Unhealthy for Sensitive Groups	Orange
151 to 200	Unhealthy	Red
201 to 300	Very Unhealthy	Purple
301 to 500	Hazardous	Maroon

- **carbon monoxide** (CO) – colorless, odorless deadly gas that, in humans and other animals, binds to hemoglobin, thereby blocking oxygen uptake by red blood cells; automobile emissions are the greatest source, but recent trends in automobile manufacturing have reduced CO emissions in the U.S.
- **lead** – gasoline with lead additives performs better, but combustion processes add lead-containing pollutants to the atmosphere; numerous human ailments are associated with "lead" poisoning, such as central nervous system impairment; lead-containing gasoline was phased out (in favor of "unleaded" gas) in the U.S. and other industrialized countries, thereby drastically reducing lead pollutants in the atmosphere; some developing nations still use leaded gas and have higher levels of lead-related afflictions
- **nitrogen dioxide** (NO_2) – reddish, strong-odored, reactive compound that contributes to formation of **smog** and **acid rain**; formed primarily from automobile emissions; **nitrogen monoxide** (NO), is readily converted to NO_2 in the atmosphere
- **sulfur dioxide** (SO_2) – colorless compound that contributes to acid rain, via formation of **sulfuric acid** (H_2SO_4); most SO_2 comes from coal combustion and generation of electricity
- **particulate matter** – liquid and solid particles small/light enough to stay in the atmosphere can be pollutants (e.g., soot and dust)
- **volatile organic compounds** (VOCs) – group of potentially harmful organic compounds typically produced during industrial processes; VOCs include methane and formaldehyde; a particularly significant group of VOCs, chlorofluorocarbons (CFCs), was correlated to ozone depletion and subsequent "holes" in the ozone layer, which normally protects Earth from excessive levels of UV radiation; *Montreal Protocol,* an international treaty in 1987, has subsequently reduced ozone-damaging VOCs by 95%; it may take decades to determine if ozone depletion has been effectively abated
- **ozone** (O_3) – while ozone is known to be an important component of atmosphere's higher portions, this same ozone is a pollutant at lower levels where humans typically are found; ground-level ozone production is the result of complex interactions among primary pollutants from automobile and other emissions; human health concerns involve general injury from oxidative processes when oxygen radicals come into contact with tissues; urban environments, on hot days, may commonly exceed acceptable EPA ozone levels

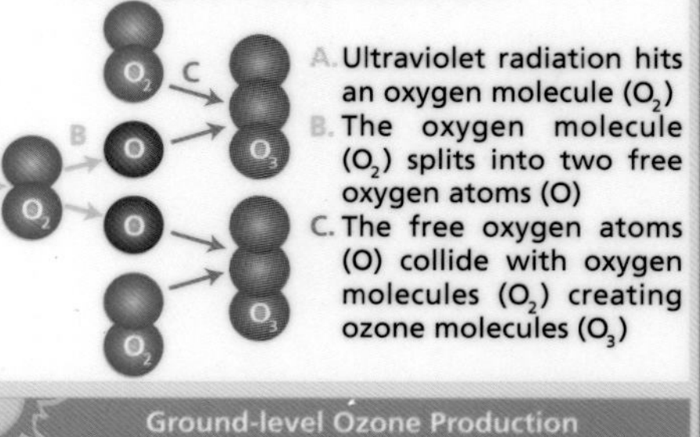

- **greenhouse gases** – in 2009, EPA declared that greenhouse gases "threaten the public health and welfare of the American people," and thus, should be regulated under the *Clean Air Act*; these gases include: carbon dioxide (CO_2), methane (CH_4), nitrous oxide (N_2O), hydrofluorocarbons (HFCs), perfluorocarbons (PFCs) and sulfur hexafluoride (SF_6); CFCs, which impact **ozone depletion,** also are considered a greenhouse gas; most anthropogenic sources of greenhouse gases come from **fossil-fuel combustion** (including gasoline, heating oil and coal) and **deforestation** (plants consume CO_2 during photosynthesis; burning to clear forests produces CO_2); collectively, these anthropogenic gases, plus water vapor (a major contributor), trap/absorb radiation energy (heat), thereby causing temperatures to rise, similar to glass of a greenhouse or windows of a car, allowing solar radiation to pass through, but blocking heat from escaping; such a phenomenon can have wide-reaching impacts

2. **Indoor Air Pollution**
   a. pollutants in the air in homes, offices and industrial workplaces are of great concern to the EPA because indoor pollutants are typically more concentrated and people spend at least 90% of their time indoors
   b. **tobacco smoke** is a deadly pollutant, with 3,000+ chemical compounds released during combustion; many health impacts have been documented; smoking is the most preventable cause of death in the world, but 95% of people live in areas lacking laws to protect them; in 2009, the World Health Organization (WHO) estimated that smoking kills 5 million people per year, including 600,000 from indirect exposure or "secondhand" smoke; U.S. laws restrict smoking to mitigate associated health risks
   c. **radon** is another dangerous pollutant (second-leading cause of lung cancer in U.S.); it is a colorless, odorless gas emanating from naturally occurring uranium decay in the ground that diffuses into overlying buildings; older homes with high radon levels can be treated; new homes can have radon-resistant materials incorporated into the general building materials
   d. **volatile organic compounds** (VOCs) and **microorganisms** (e.g., bacteria, fungi, molds) may contribute to idiopathic (i.e., undetermined) illnesses, such as **sick building syndrome**

Land Pollution

1. Damage to land surfaces and soil by human activities
   a. **urbanization** – as human populations expanded, large urban areas resulted; cities and roads have greatly altered the terrain in areas all over the world; new urban planning attempts to mitigate land damage, based on lessons learned from past mistakes; new forms of pollution are associated with urban development, such as **light** and **noise pollution**—both of which can have negative impacts on human health and welfare
   b. **soil contamination** – use of pesticides (herbicides, insecticides, fungicides, etc.) has been well documented in land pollution; toxic defoliants have been developed that more easily break down through bacterial actions (as opposed to Agent Orange); DDT, an insecticide at one time used extensively but found to biologically accumulate in terrestrial and aquatic food chains, was banned in some countries (e.g., U.S., Great Britain) because of harmful effects in top carnivores (e.g., ospreys, pelicans)
   c. **mining** – such activities can release toxic compounds, in addition to the physical alterations of the land
   d. **acid rain** – acids can form as secondary pollutants in the air, which mixes with precipitation and moves down to land (and water systems), thereby damaging vegetation (including crops) and other wildlife
   e. **waste disposal solids** – humans generate a tremendous amount of organic and non-organic wastes (both solid and liquid); such solid waste sources include residential, commercial/municipal, industrial (may or may not be hazardous substances, such as radioactive compounds), agricultural and treatment plants; waste management practices typically use two primary methods to dispose of solid wastes: **landfills** and **incineration**—both of which can be utilized beneficially for energy use; landfills generate methane gas, and incineration heat can be used to generate electricity; recycling and reusing, whenever possible, can help mitigate waste disposal problem—but a rapidly increasing human population concomitantly fuels a continually increasing waste disposal problem

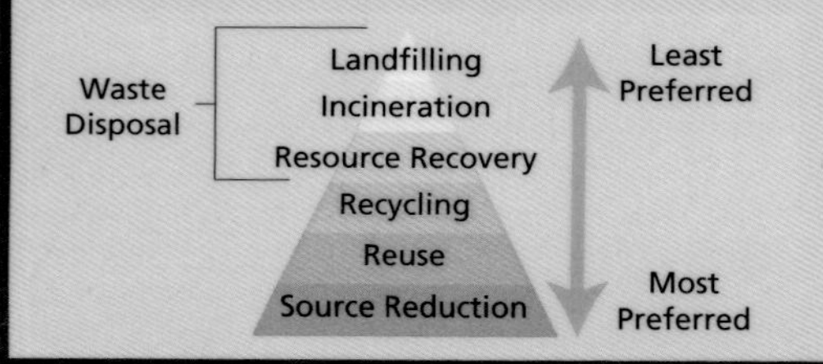

   f. **burial** – in most of the world, humans are buried upon their death; this can alter soil erosion patterns (by loosening soil) and pollute the soil (toxic decomposing compounds can leach into soil)

Water Pollution: Substances in water that potentially harm humans and the environment

1. **Freshwater Pollution**
   a. **point source** – specific origin points of pollutants can be traced back to specific sources (e.g., discharge from sewage plants or factories); in some cases, may still be difficult to regulate
   b. **non-point source** – specific origin points of pollutants are difficult to determine because of diffuse nature (e.g., large areas) of contaminate source(s), such as fertilizers and pesticides used in agriculture, and even on residential lawns
   c. **forms of water pollution**
      - **pathogens** – numerous biological agents (e.g., viruses, bacteria, protists, parasitic worm larvae) survive in surface waters; about 15% of human population does not have access to safe drinking water, and several million people die every year as a direct result
      - **nutrients** – in freshwater systems, phosphorus is generally a limited nutrient; thus, when phosphorus (and, to a lesser extent, nitrogen) enters a freshwater system, plant and algal growth occur quickly; if too many nutrients are available, vegetation may grow "too" quickly, soon exceeding carrying capacity of local ecosystem; this is called **eutrophication,** and can quickly lead to a collapse of the ecosystem as excess vegetation dies, followed by hypoxia or anoxia from bacterial decomposition; lack of available oxygen, in turn, can lead to massive fish deaths in the system (typically a lake, in freshwater); reducing fertilizer use and runoff from farms, lawns, golf courses and sewage is the best strategy
      - **synthetic chemicals** – many toxic chemicals synthesized for a variety of purposes end up in our water supplies (e.g., DDT, dioxins, PCBs, mercury, lead); toxicological symptoms in humans and wildlife vary widely, as chemicals can act as mutagens, carcinogens, allergens, eurotoxins and endocrine disrupters; once identified, goal "should" be to reduce the release of such compounds, or to look for alternatives with less impact on health/environment
      - **sediment** – many natural surface waters have relatively low sediment levels in water column; farming and deforestation activities, and subsequent erosion, can pollute clear-water systems, thereby altering these ecosystems
      - **thermal** – human activities can alter temperature of bodies of water, such as facilities that use water to cool nuclear or industrial processes; output of warm water can have profound impacts beyond discharge points, such as altering species composition or lowering oxygen content (warm waters hold less oxygen); release of water from large reservoirs may also introduce cold water into warmer river systems
   d. **groundwater pollution** – contamination of groundwater can occur relatively easily through percolation of water through contaminated soils in aquifer recharge areas; thus, it is a very serious concern in many areas of the world, as underground storage, such as septic tanks, storage containers of oil, gas, and industrial chemicals all are major contributors to groundwater pollution; a major role of the EPA is to locate and enforce cleanups and repair of hundreds of thousands of these tanks (and sometimes direct, "tankless" burial of toxic substances) in order to mitigate any further environmental damage

2. **Marine Pollution**

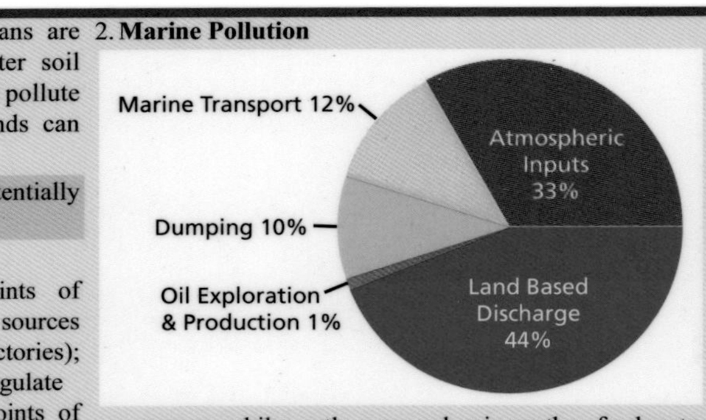

   a. oceans, while vastly more voluminous than freshwater systems, are still susceptible to significant pollution; point sources of pollution can sometimes be identified, but pollution is much more commonly from non-point sources; in many cases, humans have used (and still use) the oceans as a "dumping" ground for our wastes; specific sources of marine pollution include:
      - **land-based discharge** – coastal communities around the world (including the U.S.) discharge sewage water (treated and untreated) directly into the oceans; for example, in 2008 along the southeastern coast of Florida, at least 300 million gallons of treated sewage water were pumped into the Atlantic Ocean daily, in most cases, directly into areas with coral reefs; current legislation is attempting to resolve this problem, but complete stoppage of this pumping is still decades away, as economically viable alternatives are currently unavailable; eutrophication can also occur in the oceans, as nitrogenous wastes fuel local and sometimes regional algal blooms; **ocean dead zones,** which can happen naturally, have increased in occurrence in recent decades; evidence shows that eutrophication from fertilizers, **storm-water runoff,** and sewage (both unintentional and intentional) are major factors in their growing prevalence, in addition to naturally low oceanic currents (e.g., Gulf of Mexico) that do little to disperse the pollutants; toxic compounds found in freshwater systems can also accumulate in the oceans, albeit not as quickly, but nonetheless over time can reach critical concentrations in local ecosystems through **biomagnification**
      - **dumping** – trash and garbage, of all kinds, have been dumped into the oceans; some types of debris, such as plastics, have been well documented to have serious negative impacts on the oceans; for example, necropsies of sea turtles show that most have plastic in their digestive tracts, probably having mistaken the plastic for jellyfish prey; discarded or damaged fishing lines/nets also are serious threats to sea life
      - **atmospheric inputs** – carbon dioxide from the atmosphere readily dissolves in seawater and reacts to form carbonic acid (i.e., $CO_2 + H_2O \rightarrow H_2CO_3$); the oceans absorb about 22 million tons of carbon dioxide daily; thus, the pH of the oceans is moving downward—as carbon dioxide levels rise in the atmosphere—this trend is called ***ocean acidification***; scientists estimate that ocean chemistry is changing 100 times faster in industrial age than in previous 650,000 years; such changes are having profound effects, including erosion of coral reefs (i.e., calcium carbonate skeletons of corals dissolve in low pH conditions) and other shelled organisms critical to marine food chains; also, **acidosis** of tissues may severely impact health of fish, which could have environmental/economic impacts
      - **marine transport** – using oceans to transport can be a significant source of pollution; accidental cargo leaks/spills, such as oil from *Exxon Valdez* in 1989, can have long-term effects on ecosystems; regulations have significantly reduced risk of pollution resulting from marine transportation
      - **oil/gas exploration & production** – many vast oil reserves are located under oceans; exploiting these fossil-fuel reserves can greatly increase risks of this type of point-source pollution